Extreme Weather

TORNADOES

DiscoverRoo
An Imprint of Pop!
popbooksonline.com

Martha London

abdobooks.com

Published by Pop!, a division of ABDO, PO Box 398166, Minneapolis, Minnesota 55439.

Printed in the United States of America, North Mankato, Minnesota.

052019
092019

THIS BOOK CONTAINS RECYCLED MATERIALS

Cover Photo: Jason Persoff Stormdoctor/Alamy
Interior Photos: Jason Persoff Stormdoctor/Alamy, 1; Shutterstock Images, 5, 6, 7, 29 (sign); National Weather Service Wilmington, NC, 8; iStockphoto, 9, 11, 12, 13, 16–17, 19, 25, 27, 29 (tornadoes), 29 (checkmarks) 29 (house), 29 (flashlight), 29 (window), 29 (person), 29 (chair), 29 (television), 30; Bjorn Svensson/Alamy, 14; Jason Reed/Reuters/Newscom, 15; Jim Reed/Science Source, 20, 31; Topical Press Agency/Hulton Archive/Getty Images, 21; Kim Johnson Flodin/AP Images, 22; RosalreneBetancourt 9/Alamy, 26; FEMA/Alamy, 28

Editor: Connor Stratton
Series Designer: Sophie Geister-Jones

Library of Congress Control Number: 2018965060

Publisher's Cataloging-in-Publication Data

Names: London, Martha, author.

Title: Tornadoes / by Martha London.

Description: Minneapolis, Minnesota : Pop!, 2020 | Series: Extreme weather | Includes online resources and index.

Identifiers: ISBN 9781532163968 (lib. bdg.) | ISBN 9781532165405 (ebook)

Subjects: LCSH: Cyclones--Juvenile literature. | Rainstorms--Juvenile literature. | Severe storms--Juvenile literature. | Natural disasters--Juvenile literature.

Classification: DDC 363.34--dc23

WELCOME TO DiscoverRoo!

Pop open this book and you'll find QR codes loaded with information, so you can learn even more!

Scan this code* and others like it while you read, or visit the website below to make this book pop!

popbooksonline.com/tornadoes

*Scanning QR codes requires a web-enabled smart device with a QR code reader app and a camera.

TABLE OF CONTENTS

CHAPTER 1

HOW TORNADOES FORM

A tornado is a tower of spinning wind. Powerful thunderstorms create tornadoes. Inside these storms, warm **gusts** of wind rise. These gusts are called **updrafts**. The gusts swirl inside

WATCH A VIDEO HERE!

the storm. Sometimes, these swirling gusts create a funnel-shaped cloud.

Thunderstorms that can make tornadoes are called supercell thunderstorms.

At the same time, cold gusts of wind sink. These gusts are called **downdrafts**. The cold downdrafts wrap around the funnel-shaped cloud. They force the cloud downward. If the cloud touches the ground, it is a tornado.

DID YOU KNOW?

Tornadoes that form over lakes are called waterspouts.

A funnel cloud reaches toward the ground.

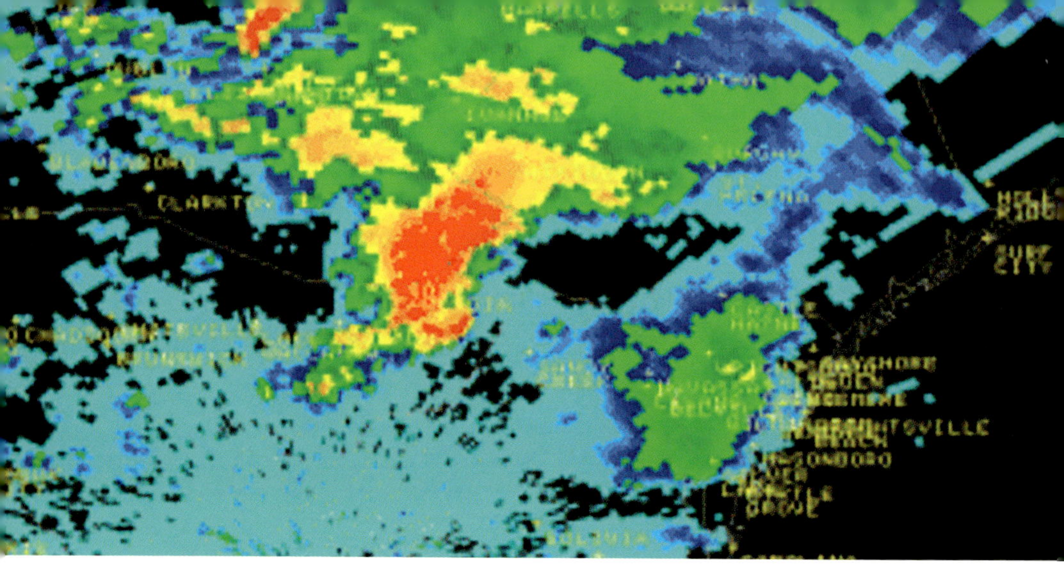

Radar uses colors to show different types of weather. For example, red means heavy rain.

Meteorologists study radar to **predict** tornadoes. Radar shows what the weather is like. Radar also shows where that weather might be headed. Meteorologists look for strong thunderstorms. These places are likely to have tornadoes.

But scientists do not understand everything about tornadoes. They do not know why some thunderstorms make tornadoes and others do not. Scientists also do not know how to predict how powerful a tornado will become.

CHAPTER 2

WHERE TORNADOES FORM

Most tornadoes happen in the United States. One area in the country receives the most tornadoes. This area is called Tornado Alley.

DID YOU KNOW?

The United States gets more than 1,000 tornadoes a year. Most last only a few minutes.

Kansas is one of the states in Tornado Alley.

LEARN MORE HERE!

The **geography** of Tornado Alley causes two air systems to meet. Dry, cool air travels south from the Rocky Mountains. Warm, wet air travels north from the Gulf Coast. When warm and cold air meet, storms can form. These storms can make tornadoes.

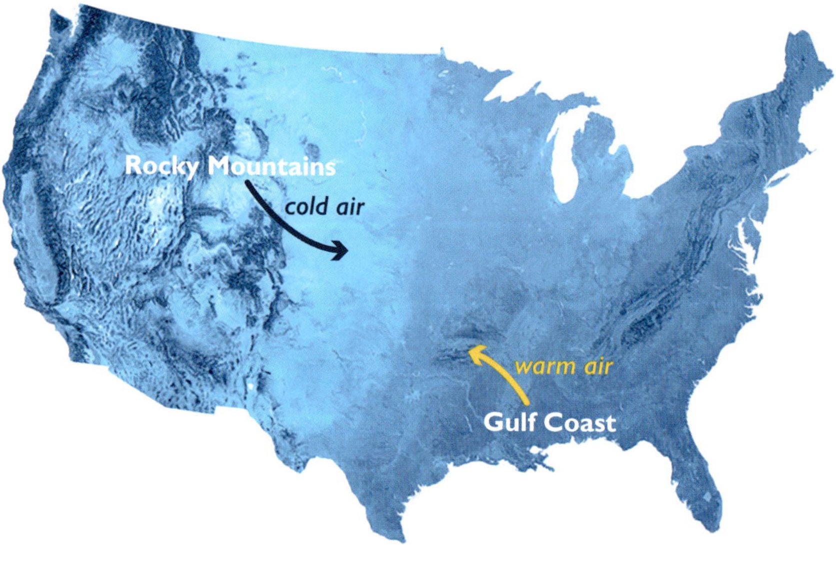

The heights of the Rocky Mountains make the air cool.

A tornado forms over the Indian Ocean.

Tornadoes also happen outside of Tornado Alley. They can hit many other parts of North America. Europe, southern Asia, and Australia get many tornadoes too.

DID YOU KNOW?

In the United States, most tornadoes happen in May and June.

In 2015, a tornado blew the roof off a house in Australia.

TORNADO ALLEY, OLD AND NEW

In the past,

Tornado Alley has

mainly included

the middle of the

United States.

But scientists have

noticed more tornadoes in

other parts of the country.

These scientists believe Tornado

Alley is getting bigger.

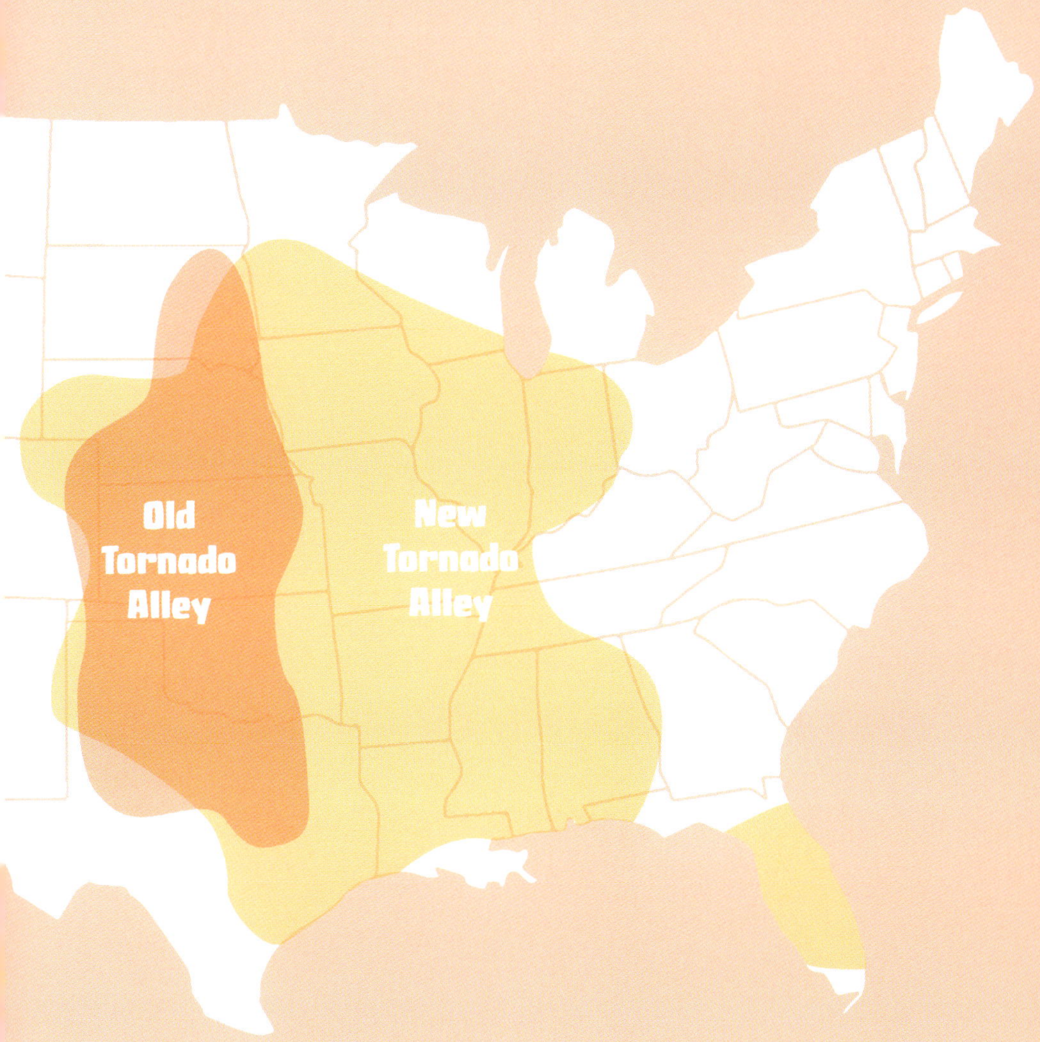
Old Tornado Alley
New Tornado Alley

CHAPTER 3

DAMAGE

Strong tornadoes can cause major damage. Their spinning winds rip up everything in their path. Tornado winds can move as fast as a speeding train. These winds can pull roofs off houses.

LEARN MORE HERE!

A tornado can flatten buildings
and rip up trees.

A tornado flung wood through this car's windshield.

Flying **debris** causes most tornado injuries. Tornadoes can pick up and throw objects, such as cars and cows.

DID YOU KNOW?

Tornadoes can pick up hundreds of frogs at once. Then, the sky "rains" frogs.

THE TRI-STATE TORNADO

In 1925, a tornado touched down in Missouri. The tornado went through Missouri, Illinois, and Indiana. The tornado traveled 219 miles. It lasted three and a half hours. It was the fastest and longest-lasting tornado in history.

Debris can even travel with the tornado as it moves. One tornado carried an object for 150 miles.

In 2013, a tornado cut a path of destruction through a city in Oklahoma.

Every tornado is different. Some tornadoes travel only a few yards. Others cause damage for several miles. Their width varies too. The biggest tornadoes can be a mile wide. But bigger tornadoes are not always more damaging. Sometimes, many tornadoes hit an area around the same time. Groups of tornadoes are the most dangerous.

DID YOU KNOW?

In 1974, 147 tornadoes formed across 13 states over the course of just two days.

CHAPTER 4

STAYING SAFE

Alerts warn people about tornadoes. One alert is a tornado watch. A watch means a tornado is possible. Another alert is a tornado warning. It means a tornado has been seen.

COMPLETE AN ACTIVITY HERE!

Meteorologists use radar devices to predict tornadoes.

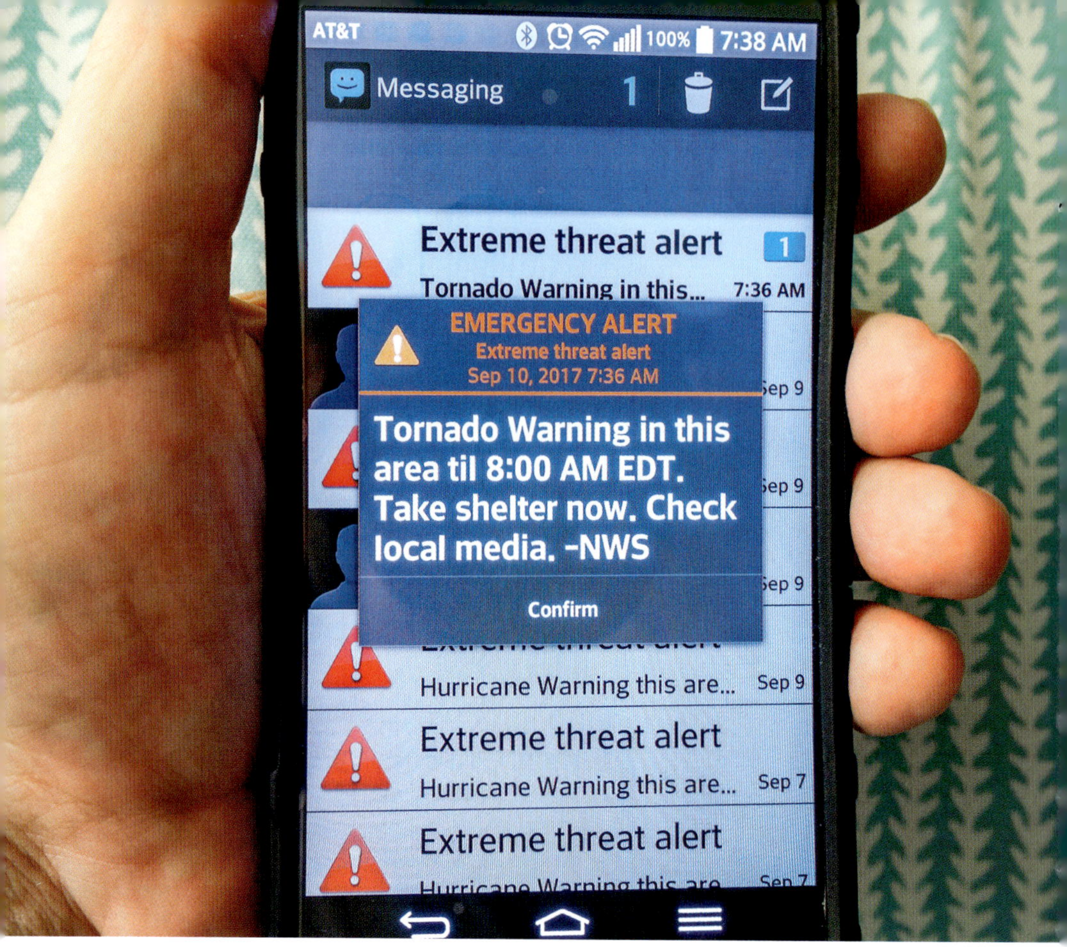

Tornado alerts are often sent to people's phones.

Some places have tornado alert systems. The systems often use sirens to warn people when a tornado is coming.

To stay safe, people rush to find shelter. They often go to the lowest room of their house. They may hide in a basement.

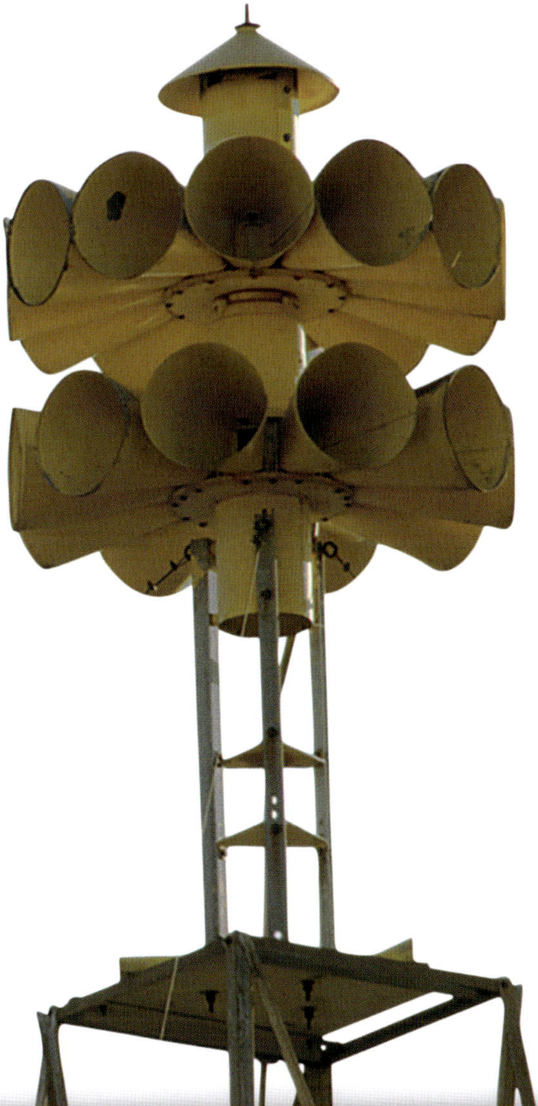

DID YOU KNOW?

After a warning, people often have only a few minutes until the tornado comes.

Girls come up from a storm cellar after a tornado.

People also try to stay near a house's center. There are more walls to stop **debris**. People may use a storm cellar too. This space below ground protects people from storms. People have many ways to stay safe during a tornado.

Tornado Safety Checklist

- ✓ Keep a flashlight in a safe place.
- ✓ Find shelter as soon as you can.
- ✓ Stay away from windows during the tornado.
- ✓ Cover your head and neck with your arms.
- ✓ Protect yourself with blankets and furniture.
- ✓ Listen to the news for weather updates.

MAKING CONNECTIONS

TEXT-TO-SELF

Are tornadoes common where you live? Why or why not?

TEXT-TO-TEXT

Have you read books about other kinds of storms? What do they have in common with tornadoes? How are they different?

TEXT-TO-WORLD

Tornadoes create flying debris. Why does going to a basement help people stay safe?

GLOSSARY

alert – a warning of danger.

debris – the remains of something broken.

downdraft – a current of air moving downward.

geography – the features of Earth's surface in a specific place.

gust – a fast, strong rush of wind.

meteorologist – a scientist who studies and predicts the weather.

predict – to say what will happen in the future.

updraft – a current of air moving upward.

INDEX

ONLINE RESOURCES

popbooksonline.com

Scan this code* and others like it while you read, or visit the website below to make this book pop!

popbooksonline.com/tornadoes

*Scanning QR codes requires a web-enabled smart device with a QR code reader app and a camera.